Lyman Abbott

A Study in Human Nature

Lyman Abbott

A Study in Human Nature

ISBN/EAN: 9783337365882

Printed in Europe, USA, Canada, Australia, Japan

Cover: Foto ©berggeist007 / pixelio.de

More available books at **www.hansebooks.com**

A STUDY

IN

HUMAN NATURE.

BY

LYMAN ABBOTT, D.D.

———— ·•·•· ————

NEW YORK:
CHAUTAUQUA PRESS,
C. L. S. C. Department.
1886.

The required books of the C. L. S. C. are recommended by a Council of six. It must, however, be understood that recommendation does not involve an approval by the Council, or by any member of it, of every principle or doctrine contained in the book recommended.

PREFACE.

THE object of this little book is purely practical. It is written to aid parents, teachers, and pastors, in their work of character-building; incidentally, too, to aid each individual to build himself. It grew out of a practical need, and was written wholly with a practical end in view.

Some years ago Dr. J. H. Vincent designed, as a part of his Chautauqua University, a Chautauqua School of Theology. Its object was not to supersede the thorough courses of biblical and theological study pursued in the seminaries, but to supplement them; to aid pastors in pursuing their studies after they had already entered on their parish work, and to enable laymen and others, who were engaged in ministerial or *quasi* ministerial labor, to equip themselves more thoroughly for their work. He proposed to incorporate in the curriculum of this Chautauqua School of Theology a "Department of Human Nature," the object of which should be to aid the student in studying man, individually and socially; human nature in history, in fiction, in the parish, and in society, thus enabling him to deal more wisely, because more truly scientifically, with the problems of individual and social life. Dr. Vincent asked me to take charge of this department, to create and to cultivate it. With much misgiving, I undertook the task; moved thereto partly by a warm personal affection and esteem for Dr. Vincent, partly by a great respect for the work which he is doing, and partly by a special interest in this particular department.

But no sooner had correspondence been opened with the students who desired to enter on this study, than we found ourselves confronted with an unexpected difficulty. There was no analysis of Human Nature which c.uld be prescribed as a basis for our proposed course. The experiment of recommending several treatises, and leaving the students to make their own analysis, was not successful, and I was thus compelled to prepare an introduction to our study before we could prosecute it. Hence this treatise, the product of studies pursued as recreation for many years, but of composition completed in a few months.

Having once undertaken to write at all, I have endeavored to prepare this Study in Human Nature, in a form so practical, so simple, and so broad, that it might be a help to every mother who desires to study the nature of her child, every teacher who wishes to study the nature of his pupil, every pastor who aims to study the character either of his parish, or of a single parishioner. All scholastic subtleties, all doubtful disputations between different schools, all technical terms, I have carefully avoided. My aim is not to expound a system of philosophy, but to incite the reader to a study of Human Nature, and to help him in pursuing it.

Mental science has fallen under a popular ban. It is thought to be a hopeless plowing of a barren soil. But the sublimest work of God is man, and there can be no worthier object of devout study than him whom God has made after his own image ; and surely no object about which we are more concerned to know, whether we regard our own welfare or the well-being of our fellow-men. To all who love their fellow-men, and desire to know and serve them better, this little attempt to aid them in that knowledge and service is dedicated by

THE AUTHOR.

CORNWALL-ON HUDSON, N. Y.

CONTENTS.

A STUDY IN HUMAN NATURE.

CHAPTER I.

THE NECESSITY OF THE STUDY.

" KNOW thyself" was an ancient Greek apothegm. Lord Beaconsfield, in his famous address before the University of Edinburgh, declared that the fundamental conditions of success in life were two—knowing one's self, and knowing the needs of one's age or epoch. But of all knowledge, self-knowledge is the rarest; perhaps, also, the most difficult to attain. It is only recently that physiology has become a study in our schools. Until within a few years all knowledge of the body was thought to be a specialty belonging only to the doctors. Even to-day mental science—the organism and operation of the mind—is not studied in our schools. This is left to the higher classes in our colleges, and studied there as an abstract, not as a practical, science. Every man ought to know his own nature; his bodily strength and weakness ; his mental strength and weakness ; his moral strength and weakness. A knowledge, concrete, not abstract, practical, not theoretical, of human nature, is essential to the best and truest success in life—to health, to development, to usefulness.

1. No man can keep either mind or body in health unless he knows what his mind and body are. He cannot keep himself in order unless he knows how he is constituted. The body is a wonderfully delicate machine. It is placed in a world where there are many influences at work destructive of it. There is poison in food, in water, in air; there is

"death in the pot." There is evil in excess; there is evil in scant measure. Men suffer from too much air and from too little; from over-feeding and from under-feeding; from excessive sleep and from too little sleep; from too violent exercise and from too little exercise. We must know not only what, but also how much, our bodies need. The specialists aver that most men have a streak of insanity in them. A thoroughly sane mind is as rare as a thoroughly healthy body. To keep the mind well balanced, to preserve it in good order, to enable it to work clearly, quickly, efficiently, regularly, requires a knowledge of the mind and of the conditions of mental health. The ministers assure us that all men are diseased morally. Life abundantly bears out their assertion. No man is perfectly healthy, morally; for perfect health is a perfect balance of all the moral powers. Every faculty has its own disease. The conscience may become cruel—witness the Inquisition. Religion may become superstition—witness the history of all pagan and some forms of the Christian religion. Love may become sentiment—witness the story of many a child ruined by the false love of a doting mother. And observe that every man's body, mind, and spirit is distinct from every other man's. Its conditions of health are peculiar. What is one man's meat is another man's poison. One man needs cereals, another meat; one man needs to read more fiction, another needs to abandon it altogether. One man needs to cultivate his reverence, another his conscience, a third his sympathy. To produce, to cultivate, to maintain health of body, mind, and spirit, every man has need to know his own nature, the laws of his own being, the condition of his own health.

2. Self-knowledge is equally indispensable to growth, education, development. Some of the Hebrew scholars tell us that the familiar text in Proverbs about child-training should read: Train up a child in *his own way*, that is, according to the bent of his natural genius, and when he is old he will not

depart from it. Whether this is sound exegesis or not, it is certainly sound philosophy. We must know the nature of what we would develop. We must understand what it is before we begin to shape and fashion it for its future. Self-knowledge is the condition of self-culture. Are you deficient in imagination? You must both know that fact, and what are the methods of developing imagination, or you cannot grow symmetrically. Has God endowed your boy with qualities which fit him for the merchant? you only waste your time, and destroy his usefulness, by trying to make a minister of him. If Martin Luther's father could have had his way we should have had no Reformation, or a very different one; for he wanted to make a lawyer of Martin. History is full of instances of men who knew their own nature better than their parents did, and so came to something in spite of parental blunders; and still fuller of instances of men who neither knew themselves nor were understood by their parents, and so came to nothing. The best seed will produce fruit only in the hands of one who knows what it is, and therefore what soil and cultivation it requires. Moral development requires moral self-knowledge. There is not one specific for all sins. Christ is not the world's medicine, but the world's physician; and his prescriptions are various for various disorders. To grow in holiness is to grow in healthiness; and this requires a knowledge of your own nature, that you may know what needs feeding and what needs pruning. Some men are weak through lack of self-esteem, and some men through too much. Some men pay too much attention to other people's opinion, and some men too little. Some men pay a blind reverence too easily, and some scarcely know what reverence means. Each nature requires its own education. The training which will help the man of undue self-esteem, will hurt the man who has too little. A chief end of life is to grow aright; and no man can grow aright unless he understands the principles of his own nature.

1*

3. For the same reasons a knowledge of the principles of human nature is essential to the highest and best usefulness. A knowledge of human nature is the first condition of the successful conduct of life. Every business man, lawyer, doctor, statesman, needs it. If a man should attempt to farm without any knowledge of seeds and soils, or to mine without any knowledge of metals, he would be sure to fail; how can he succeed in dealing with men if he knows nothing about human nature. The merchant needs this knowledge to select his salesmen; the salesman to sell his goods; the doctor to secure and retain the confidence of his patients; the statesman to adapt his laws and policies to men as they are; the editor to provide intellectual food that actual readers will read and profit by. All successful men have a knowledge of human nature. Sometimes they have acquired it empirically, not scientifically; that is, they have picked it up by their dealings among men, not by a careful study of principles; but in one way or the other they have got it.

4. This knowledge is essential to the well-being of the family. Every girl ought to be taught the general principles of human nature, for it is probable that she will be a mother, and she needs this knowledge to know how to care for and to train her children. One of the great causes of domestic infelicities, quarrels, and divorces is ignorance of human nature. The husband and wife do not know either themselves or each other; they do not know how to correct their own faults or the faults they see in each other. If they did, they would have hope of curing the present evil; and hope would give patience; and patience would prevent bickerings, and strife, and separation.

5. Especially is this knowledge of human nature necessary to all men whose professional duty it is to train or instruct others. The teacher needs it. It is more necessary to him than a knowledge of Greek, or Latin, or mathematics. He must know the minds which he is to mold and the laws by

which they are to be molded. There have been many scholars greater than Thomas Arnold, of Rugby, but never a greater teacher, because he knew so thoroughly well boy-nature and its laws. The minister needs it. It is more necessary to him than a knowledge of Hebrew or theology. He must know the natures he is to cure, and how to cure them. He must know the pathology of pride, vanity, covet-ousness, ambition, passion, if he would either mend the man-ners or change the lives of his congregation.

The object of this little treatise is to afford some help to ministers, teachers, parents, and men and women generally, who wish to understand the general principles of human nature, and to aid them in a study of men, and women, and children, for the purpose of protecting them from temptation, developing them, and building them up into a Christian manhood and womanhood—perfect men in Christ Jesus. Its object is wholly practical; its style will be as simple and plain as I can make it.

CHAPTER II.

A PRELIMINARY QUESTION.

IT has been greatly discussed among philosophers whether the mind is simple or complex; whether it is one and individual, or made up of various distinct powers and faculties; whether one and the same power imagines, reasons, remembers, feels, or whether there are distinct powers, of which one imagines, another reasons, a third remembers, a fourth feels. Let me first get this question clearly before the reader's mind.

Man is equipped with various senses, each of which has its own peculiar function. It can perform that function, and no other. The ear can hear, but it cannot see; the eye can see, but it cannot taste; the palate can taste, but it cannot smell. The body is composite. It is made up of different organs or faculties. The whole man is an orchestra; each organ is a single instrument. If that is broken, or gets out of tune, no other can take its place. Now some persons suppose that the mind is similarly a composite; that it is made up of a variety of faculties and powers; that there is one power or faculty which reasons, another which compares, a third which remembers or recalls, a fourth which imagines, etc. Those who hold this opinion, however, are not agreed as to how many mental faculties or powers there are. Some suppose there are very few, others that there are very many. A very common classification or division of the mind is into three powers or classes of powers: the reason, the sensibilities or feelings, and the will. Others divide these generic classes again into a great variety of reasoning and feeling powers, each confined to its own exercise or function, as the ear to hearing and the eye to seeing.

Other thinkers suppose that there is no such division; that the mind is not made up of a variety of organs at all, that it is simple and indivisible. The mouth is so formed that it can perform two very different functions—it can eat and it can speak. There are not two organs, one an eating and the other a speaking organ, but one organ which now eats, now speaks. So some scholars suppose that there is one mind or soul which is absolutely indivisible, but which exerts itself in different ways at different times: it sometimes remembers, sometimes imagines, sometimes loves, sometimes hates, sometimes reasons, sometimes chooses; but it is always the same power which remembers, imagines, hates, loves, reasons, and chooses.

Now, this is a question on which no absolute conclusion can be reached. We cannot analyze the mind as we can analyze a substance in a laboratory, and see what are its constituent parts, and determine whether it has any parts, or is a simple substance. We have only two methods of judging about the mind, and neither of these methods gives us any conclusive answer to the question whether the mind is simple or complex. We can observe the operation of other men's minds by studying its results in action, or in speech which is a kind of action; and we can study the action of our own · minds by looking within and seeing what our own thoughts and feelings are. But, in both these cases, we study only the operations, not the mind itself; and neither a study of the results of mental operations in the actions of men and women about us, nor a study of our own mental operations by looking within and trying to ascertain of our self-consciousness how we think and feel, throw any important light on the question whether the mind itself is simple or complex. All that we *know* about the mind is its operation; all else is theory.

There are some metaphysical and abstract arguments for the opinion that the mind, the I within, that controls the body, what the Germans call the *ego*—which is Latin for I—is simple, not complex; that is, one power operating in different ways and doing different things. I am myself inclined to think that the better opinion; but it is not necessary here to go into this question at all, for what we are going to study is not the mind itself, but human nature, that is, the *operations* of the mind. And there is no doubt that the operations of the mind are complex. There may be, I am inclined to think there is, but one power, which perceives and thinks and feels and wills; but perceiving and thinking and feeling and willing are very different actions, and it is only with the actions that we have to do.

In this book, then, I speak habitually of the different faculties or powers of the mind. The reader must understand that I do not mean by this phraseology to imply that the mind itself is divided into different powers, each with its own peculiar function. But in order to study mental phenomena we must form some classification of them, and must analyze them under different divisions and subdivisions. When, for example, we speak of the faculty of comparison, we do not mean that the mind has one power which compares and observes the relation of things, and can do this and nothing else; but we mean that the mind has a power of observing the relations of things, and this power we call the faculty of comparison. In the same way we might say that the mouth has a faculty of eating and a faculty of speaking and a faculty of singing, without meaning that there are in the mouth three sets of organs, of which one eats, another speaks, and a third sings.

I do not wish to leave the impression that the question whether the mind is simple or complex is one of no special consequence; only that it is not necessary for us to determine it in order to our present plan of study. It has an important, though perhaps rather indirect, moral bearing. That bearing may be briefly mentioned here. If a man has

a good ear, but poor eye-sight, we cannot say that his faculties are either good or bad; one faculty is good, the other is
bad. Now, if man is made up of a bundle of faculties, some
of which are good and the others bad; if, for example, his
conscience is strong, but his love and sympathy are weak, we
cannot say of him that he is either good or bad; part of him
is good and part of him is bad. If, on the other hand, he is
a unit, and conscience is simply the man acting in one direction and love is the man acting in another, then we cannot
truly say of him that he is good until all his actions are con-
formed to the divine standard. If, for example, a carpenter
has a box of tools containing a chisel of soft iron and very
dull, and a plane of finely tempered steel, very sharp, and we
ask him what sort of a set of tools he has, he would reply,
some of the tools are good and some are poor. But if an
apprentice has learned to drive a nail without splitting the
wood, but he cannot yet saw a straight line, there is no sense
in which we can say he is a good carpenter. He is not a
good carpenter until he has learned to do all carpentering
operations at least reasonably well. This illustration will
give a hint of the argument for the simplicity of the mind,
of which I spoke above. When I yield to my anger and
speak a bitter word, I am conscious that I have done wrong:
not that some thing in me has done wrong, but that the whole
I has sinned; and this, perhaps, is what James means when
he says that he who has kept the whole law and yet offends
in one point is guilty of all. It is the soul that sins, not a
faculty in the soul. Thus there is a reason in our consciousness of sin for believing that the soul or mind—the *ego*, the
I—is a unit, not complex or composite. In this book, however,
in speaking of mental and moral action, I shall, for convenience'
sake, speak of mental faculties, meaning thereby not separate
powers, but separate activities of the same power working in
different ways.

CHAPTER III.

TRUE AND FALSE MATERIALISM.

It is common, even in the pulpit, to hear the phrase, "Man has a soul;" and it is scarcely possible to avoid embodying this same thought sometimes in the phrase "man's soul," which is only an abbreviation. This phrase, however, expresses a falsehood. It is not true that man has a soul. Man is a soul. It would be more accurate to say that man has a body. We may say that the body has a soul, or that the soul has a body; as we may say that the ship has a captain, or the captain has a ship; but we ought never to forget that the true man is the mental and spiritual; the body is only the instrument which the mental and the spiritual uses.

Still more accurately, however, man, as we see him and have to do with him in this life, is composed, in Paul's language, of body, soul, and spirit. The distinction between these three we must consider hereafter. Here it must be enough to say : 1. That the body is purely physical, as much so as a tree; that it is composed of certain well-known physical elements, and subject to physical laws. 2. That the mind or soul (in Latin the *anima*, in Greek the *pseuche*) is that which sees, feels, thinks, and that it is analogous to that which controls the body in the animals, though in man possessing powers vastly superior to those observed in any mere animal. 3. That the spirit (in Latin *spiritus*, in Greek *pneuma*) is that which deals with the invisible, believes, reverences, distinguishes between right and wrong, and that there is nothing analogous to it in the animal creation. The body links us to the earth, the mind to the animal creation, the spirit to God.

To understand human nature we must understand the relation which the mind and spirit, that is, the invisible part of man, has to the body, that is, to the physical or material part.

1. It is now well established as a scientific fact that every mental and moral act employs some physical agency and makes a draft upon the physical organization. In fact, every mental action is also partly a material and physical action. We know, for instance, that we see by means of a physical organ, the eye; we hear by means of a physical organ, the ear. The eye does not, however, see ; for if the nerve which connects the eye with the brain be cut, though the picture is perfectly painted on the retina of the eye, the person sees nothing. So the ear does not hear; for if the nerve which connects the drum of the ear with the brain be cut, the person hears nothing. The seeing and the hearing take place within us, and the eye and the ear are only the physical instruments by which they are facilitated. The eye no more sees than the telescope ; the ear no more hears than the ear trumpet. But both are necessary instruments to seeing and hearing. For aught we know, however, both eye and ear may be destroyed as they are at death, and the power of seeing and hearing possessed by the soul may be improved, not impaired, by the loss of the instruments.

Now as the eye is the instrument of seeing, and the ear of hearing, so the brain is the instrument of thinking and feeling and imagining. Every mental and moral action employs some portion of the brain, as every act of seeing employs the eye, and every act of hearing employs the ear. Not only that, but every such action destroys a part of the brain, and a new brain tissue must be formed to take its place. Every action of the man, physical, mental, or moral, wastes some tissue. The principal physical function of life appears to be carrying off this wasted and exhausted and now useless tissue by various methods of drainage, and supplying new tissue to take its place by various methods of food supply.

In all ages of the world the use of physical organs by the mind and spirit has been recognized, not only by the philosophers, but also by the common people. The ancient Hebrews put the seat of the emotions in the bowels; hence the phrase, " bowels of mercies," as used in Scripture. This was probably because strong emotion affects the bowels. Later, for an analogous reason, because of the effect of strong feeling on the heart and circulating system, common language fixed upon the heart as the seat of the emotions. This notion still lingers in such phrases as " a warm-hearted friend," " a good-hearted fellow." But it is now well established that the real seat of both the affections and the intellect is in the brain. By this is not meant that they are located in the brain. They have no location; they are omnipresent in the body, as God is omnipresent in the universe, equally controlling all its parts. It is more accurate, therefore, to say that it is now well established that the material or physical organ of all thought and feeling is in the brain; that every mental and emotional activity employs some part of the brain ; that every such activity uses up some brain tissue, requiring, therefore, a new supply; and that, therefore, the healthful action of the mind requires a good brain, and the best action of the mind requires good digestion and good circulation, since on these depend the renewal and replenishing of the brain. .

There are various grounds for this now well-established conviction. They are all summed up in the general statement that any disease of the brain produces mental and moral disease, while, on the other hand, no disease which does not directly or indirectly affect the brain, has any power to affect the mental and moral sanity of the patient. Thus a blow on the knee which will produce excruciating pain will leave the mind clear, while a blow on the brain will produce unconsciousness. A gastric fever does not materially alter the apparent moral condition of the sick man, at least not more than might be expected from the effect on the brain of so serious a disease in the organ on which it depends for its supply. But a brain fever makes the patient delirious, and sometimes changes entirely his apparent intelligent and moral character. Thus I have known of the case of a young man, of most exemplary character, who was almost morbidly sensitive to any word or phrase of an indelicate or coarse description, who, being taken with brain fever, was so blasphemous and obscene that it was impossible for any female attendant to remain in the room with him. It was clear that the disease was physical, not moral ; it was a disease, not in the mind or spirit, but in the organ which they employed. The difference may be compared to that which would occur if a Rubinstein should sit down to play upon an old and out-of-tune piano. The discords would be due to the instrument, not to the player.

If the brain is impaired the mind is invariably affected ; if, on the other hand, the brain is uninjured, the mental and moral powers will remain unaffected, though the rest of the body may be to all intents and purposes well-nigh dead. It is true that the brain is so closely connected with the nervous system, which pervades the whole body, that any thing which impairs the nerves of the body impairs the brain, and therefore affects the mind ; but the general principle, that every other part of the body may be weakened and the mind be left comparatively unimpaired, *provided the brain is uninjured*, has had many striking illustrations in the history of great mental work achieved by chronic invalids. A very striking illustration of this is afforded by the extraordinary story of John Carter. At the age of twenty-one he fell from the branch of a tree, forty feet in height, and was taken up unconscious. Examination showed a severe injury to the spinal column, effectually disconnecting the brain from the rest of the nervous system, and depriving the body of all power of motion from the neck downward. He soon recovered consciousness, but never moved a limb again. But his brain, and with it the powers of his mind and spirit, were unimpaired. From being ungodly and ignorant, he became both devout and intelligent, a great reader, and soon learned to write, to draw, and even to paint, holding the pencil or the camel's hair brush between his teeth, enlarging or reducing the copies before him with great artistic skill and perfect success. He lived in this condition for fourteen years, his whole body from the neck downward being paralyzed and helpless, while his mind and spirit were not only uninjured, but grew brighter and clearer to the end. It was evident that the accident which had left only the head uninjured had left all the organs of thought and feeling uninjured.[*]

It is now, then, well established as an undoubted scientific fact that the mind or soul acts through organs; that these organs are in and form a part of the brain; that their healthy action depends upon the healthy condition of the organ, that is, of the brain; that any thing which impairs the health of the brain impairs the healthful action of the mind or soul, though how it affects the mind or soul itself we cannot say; that what we call mental diseases are often diseases of the organ; that the remedy for what appears to be an imperfect or evil action of the mind or soul must sometimes be applied to the organ, that is, it must be physical rather than mental or moral; and that whoever has to do with the training, education, or development of men, has a need to study the relations of the mind to its organs, and to ascertain, as far as possible, what diseases and what hinderances to development are mental and moral, and what are material or physical; and, finally, that he who would attain the highest degree of manhood must study, not only to improve his soul and spirit by

---

[*] See an interesting monograph, " The Life of John Carter," by F. J. Mills. Hurd & Houghton, 1868.

intellectual and spiritual processes, but also to care for and nourish properly his brain, that is, the organ of his mind, soul, and spirit. A healthy man is *sana mens in sana corpore*, a healthy mind in a healthy body. Well-being requires healthy organs as well as a healthy mind to use them. The physician needs often to inquire into the condition of the mind in order to prescribe intelligently for the body. The minister needs often to inquire into the condition of the body in order to prescribe intelligently for the mind and the spirit. A sleepy congregation is oftener the sexton's fault than the preacher's. Depression of spirits may be due to remorse; it may be due to a poor digestion or a diseased liver. Remedy for apparent sin may be Bible and prayer; it may be less food and a run in the open air. The teacher, the parent, the preacher, needs to study with care the condition of the body in order to deal wisely and well with the intellectual and the moral condition of those intrusted to their charge. Moral reformation and material reformation must go together. It is almost hopeless to promote temperance and godliness in our great cities so long as the population live in some wards with more persons to the square foot than are allowed in the average cemetery. The best prevention of crime is often a change of air, food, and other physical conditions. The great majority of street boys, if left in New York city, grow up to swell the number of the criminal classes. But last year the " Christian Union " sent out to Minnesota, through the Children's Aid Society, some one hundred and twenty-five children. Of these, all but five are doing well; that is, they are making good, industrious citizens. Much is due to a change in moral and intellectual circumstances, but something is also due to a change in physical circumstances.

Modern science has gone further in its investigation. It is beginning to learn that different parts of the brain perform different functions. It is now well settled that the organs of sense, of intellect, of feeling or emotion, and of will, are not the same. But these investigations are not yet completed, and it is not necessary for our purpose in this little treatise to enter upon this branch of the subject.

It is necessary, however, before closing this chapter, to note the difference between the doctrine that the mind acts through organs, and is therefore dependent for its practical results upon the health of the organ, and the doctrine that there is no mind, but that which we call mental and moral action—thought, feeling, and will—are the effects of material changes taking place within the body. This doctrine goes by the name of materialism. Among the ancients there was a class of philosophers who taught that God did not create the world, but the world created God; that is, they held that matter is eternal, and that spirit was evolved out of matter. Analogous to this is the doctrine of modern materialism; the doctrine that the body is not the instrument which the mind or soul uses, but the machine whose action produces the mind or soul, somewhat as the friction between the grindstone and the scythe produces sparks. It is unquestionably true that every mental and moral action is accompanied with a change in the brain. The materialist, observing this, has jumped to the conclusion that the change in the brain produces the mental and moral activity. This is a long jump.

1. In the first place there is no evidence whatever to warrant this conclusion. It is as if a boy seeing an organist playing on an organ should conclude that the keys of the organ moved the fingers of the player. We do know that the mental and the brain actions are contemporaneous and concomitant; but this gives us no reason to suppose that the brain action produces the mental action, or that the mental action produces the brain action. Which is the cause and which the effect we must learn in another way.

2. If the organist were an automaton, the boy would be left

in doubt whether the machinery which moved the organ was in the organ or in the man. Unless he could take one or the other to pieces, he could not tell which was the agent and which the instrument; which acted, and which was acted upon. Now we cannot look within our neighbor to see whether the brain moves the mind or the mind the brain; but we can look inside ourselves and see which moves first. We do this by self-consciousness. And this assures us that the mind operates first, and the brain and nervous system afterward. The artist is conscious that he forms in his mind a picture before his hand begins to put it upon canvas. We know that we will to reach out our hand or stretch forth our foot before we move the organ. Walking does not make us desire to go; the desire to go makes us walk. So far as we can trace mental and moral action at all within ourselves, it is clear that first comes the desire, then the will, then the action. It is very evident that the visible organs, that is, the eye and hand and ear, are the servants, not the masters; there is no reason whatever to suppose that the invisible organs, that is, the brain organs, are the masters, not the servants.

3. If the organ produces the activity, if the brain secrets thought and feeling as the liver secrets bile, as has been claimed by the materialist, there is no such thing as right and wrong. Man is a mere physical machine. His thought and feeling and will have no more moral character than the sparks of an electrical machine. Garfield was simply a good and useful machine; Guiteau was simply a bad and dangerous machine. It is true that even on this theory we might still continue to put the good machine where it would do the most good, and destroy the bad one; we might elect a Garfield to the presidency much as we would put a good time-keeper on the mantle-piece, and destroy a Guiteau, much as we would knock to pieces an infernal machine. But we could no longer approve the one and condemn the other; and in fact materialists do either actually deny that there is any

such thing as virtue and vice, or make very little of the distinction between the two. But no philosophy of man can be true which denies the most fundamental fact in human experience, the fact of oughtness, a distinction between right and wrong, the sense inherent in all men .that some things are right, honorable, praiseworthy, and that other things are wrong, dishonorable, worthy of condemnation and punishment. The family, society, citizenship, are all built on the recognition of this fundamental fact which materialists either deny or ignore.

4. If the organ produces the action there is no reason to suppose that the action will survive the organ ; if the brain feels, thinks, reasons, wills, when the brain crumbles into dust the thinking, reasoning, feeling, willing, will cease. When the fuel is burned out the fire will cease ; when the battery is exhausted the electrical current will cease. According to materialism the brain is a fire, and all mental and moral phenomena are only the heat it gives out; the brain is a galvanic battery, and all thought and feeling are only the electric current which it produces. Now we have nothing to do here with the morality of this doctrine; we are not considering its moral effect, but its reasonableness. A doctrine which has nothing whatever to support it, and has against it the almost universal instincts of mankind, is not reasonable. And the instinct of immortality is the almost universal instinct of mankind. We feel our immortality before we pass from the body, much as the bird feels conscious of the power of flight before it is fledged, or has attempted to leave the nest. We are conscious of something within which is imperishable. But if the organ produces the action, there is no such imperishable power within ; the pains of remorse do not differ from the pains of dyspepsia, nor the joys of love from those of appetite. No one can really believe this ; no one acts as though he did, not even those philosophers who imagine that they believe it.

5. Finally, if the organ produces the action, then there is no personality. There is no I that thinks, reasons, feels, acts; there is only a succession of nerve phenomena which we call thinking, reasoning, feeling, acting. If the brain is a kind of galvanic battery, and feeling and thinking are the sparks, then I am only the succession of sparks. This has been seen and acknowledged by the materialists themselves. Thus Hume, declaring that there is no such principle as self in one, goes on to affirm of mankind that "they are nothing but a bundle or collection of different perceptions which succeed one another with inconceivable rapidity, and are in a perpetual flux and movement." This is the logical conclusion of materialism, or the doctrine that the organ moves the organist, not the organist the organ; and it arouses against itself the instant testimony of our own consciousness. If there is any thing that we *know*, absolutely and positively, it is that we exist; that there is an I which perceives, feels, reasons, wills, and that is as separate and distinct from the mere succession of perceiving, feeling, reasoning, and willing, is the player is from the succession of notes which he produces on the organ. That there is both an I and a not I is perfectly clear to every one of us. The doctrine that there is no I, no self, no personal identity, can never make any greater progress among mankind as a practical doctrine than the doctrine of Berkley, that there is no external world, and that instead of real objects which we think we see, hear, touch, taste, and smell, there is only a succession of impressions, a seeing, hearing, touching, tasting, smelling; that we are all living in a dream, some in a delightful one, others in a nightmare; that life is only a kind of phantasmagoria. The one philosopher denies that there is any thing not I; I is all there is. The other denies that there is any I; what seems to be so is only a succession of physical forces. It is doubtful whether any man really believes either of these notions. And it has been necessary to point out the absurdities in-

2

volved in the doctrine of materialism only in order to make perfectly clear to the reader the distinction between true and false materialism. The true materialism teaches that the mind and spirit act always in this life through organs, and that healthy mental and moral action depends upon healthy organs. This is established by a variety of physical experiments, and is now undisputed. The false materialism teaches that the material organism produces all mental and moral phenomena; and it is without any evidence whatever to support it, is a purely abstract notion, and is contradicted by our consciousness of our own actions, by our inward sense of the distinction between right and wrong, by our instinct of immortality, and by our certainty of personal existence and identity.

CHAPTER IV.

THE TEMPERAMENTS.

FROM a very early age physiologists have recognized a characteristic difference between persons possessing the same organs, and yet manifestly possessing different qualities. These characteristic differences have been called temperaments. How far they are physical, how far mental, is a question not necessary here to discuss ; certainly it has not been determined. But that they are partly physical is unquestion-able. Various classifications have been suggested of these temperaments, no one of which is altogether satisfactory ; but there is, perhaps, none better than the one which is at once the simplest, the most common, and very ancient, into the nervous, the sanguine, the bilious, and the lymphatic. In the person of nervous temperament the nervous organism is the predominant one ; usually the head is large and finely formed, the skin fair, the complexion light, the hair fine and generally dark. Any one of these signs, however, may be wanting, and the person still possess a highly organiz d and delicate nervous system. A more certain indication of it is sensitiveness to impressions, both physical and mental, subtle and readily responsive sympathy, and quickness and alert-ness of action both in mind and body. The person of ner-vous organization is also often able to sustain an amount of labor or suffering far beyond what would be anticipated of him from his general physical condition, but always at the hazard of a sudden and sometimes an irretrievable collapse, following the expenditure of nervous force, not adequately kept up by other organs. Such a person is also liable to great fluctuation of feeling—both exaltation and depression,

dependent on the condition of the nervous system, and sometimes upon slight external circumstances acting upon it, as
the weather, food, or drink, or even social sympathy, or the
lack of it.   In the person of sanguine temperament the blood
currents are rich and strong; the whole nature is therefore
well fed; the nervous system, whatever its capacity, is habitually at its best.   Such a person has usually a rich color, often
red or reddish hair, generally a light eye, and a bounding
motion.   A surer indication is vigor in action and hopefulness
in feeling.   To act is a pleasure to the sanguine; idleness is
a vice which he cannot understand; weariness a weakness
with which he has not easily any sympathy.   And as it is a
delight to cope with difficulties, they have no terror for
him, and he carries into every exigency a hopeful spirit.   He
scarcely knows the meaning of despair.   The reader will
find a fuller description of this temperament in Campbell's
immortal· verse, which may serve a better purpose than
a more scientific description would do.   The bilious temperament is the reverse of the sanguine, and is, indeed,
rather the product of a disease than of the predominant activity of a healthy organ.   Physiologists are not agreed among
themselves as to the function of the liver or the effect or
object of bile, but unquestionably one of the chief functions
of the liver is to eliminate from the system the waste, that is,
the dead tissues after they have served their purpose, and
bile is at least in part an excretion of materials which are
decomposing and ready to be removed from the system.
When the liver fails to do its work properly, and these materials are not removed, but remain in the blood to circulate
again through the system, which they cannot feed any more
than the ashes of a burnt coal can feed the fire, the man is
said to be bilious; when they exist in the system to a large
degree he is poisoned, and if the poison cannot be removed
he is certain to die.   When as a habit of the body, very apt
to be produced by sedentary habits, or excessive or unwise

food, the liver thus fails to eliminate from the circulation matter which should be removed, the power of activity of every kind becomes impaired, exertion is difficult, thought is slow, the head is dull and stupid, small difficulties grow serious to the imagination, and the whole mood becomes both inert and melancholy. A person of this temperament is ordinarily of a sallow complexion, of dark hair, sluggish in action, and depressed in spirits. The lymphatics also share in the work of removing the effete tissues from the system. When they fail to fulfill this function, the waste material remains in the system, not, however, in the blood, but in the tissues. These add nothing to the real vigor of the man, because they are an addition of valueless and really dead tissue. Such a man is loaded down like a locomotive which should be compelled to carry in the tender its own ashes. He is likely to be obese, though not necessarily offensively so; he is certain to be sluggish and good-natured; not quick to take offense, because not quick to action of any kind; habitually content; rarely or never giving himself to work spontaneously, but only under the pressure of some motive, and always glad to relax his work and drop into idleness again. In fiction, the fat boy in Dickens's " Pickwick Papers" is a travesty on the lymphatic temperament. Mr. Bain has suggested that to this ancient classification of temperaments should be added the muscular temperament, in which the muscular system predominates, in which physical action is enjoyed for its own sake, which creates a love for field sports and athletics of all kinds, and of which probably the Roman and Grecian gladiator might be regarded as extreme types.

I have necessarily spoken of each of these temperaments as distinct from every other; in fact, all temperaments are a combination. In every man something is contributed by the nerves, the blood, the liver, the lymphatics, and the muscles; no two men were ever composed in the same way; the variations are endless. Frequent combinations are the nervous-sanguine, the nervous-bilious, the nervous-muscular, the lymphatic-sanguine, the lymphatic-bilious, and the muscular-sanguine.   As active exercise is the best method of keeping the body free from its own degenerate and wasted tissues, and assuring their elimination from the system, the muscular is rarely found in combination with either the lymphatic or the bilious, and for the same reason the bilious is rarely found in combination with the sanguine, since the life currents can never be vigorous and healthy when the body is choked with its own waste.   It is, however, certain that in any estimate of human nature, and in any study of the individual, the student must bear in mind the effect which the predominance of these temperaments or their combination— the nervous, the sanguine, the bilious, the lymphatic, and the muscular—may have upon mental and moral activity.

CHAPTER V.

· ANALYSIS OF HUMAN NATURE.

WE are now prepared to enter upon an analysis of human nature. In doing so, however, I must first again remind my readers of the object of this treatise, and of the fundamental principles already laid down.

1. The object of this treatise is not to afford an anatomical chart of the human mind. It is not to explain what are the powers or faculties of which the human soul is composed. Whether the mind is simple or compound is not the question here ; for myself, I regard it as simple ; not as, in strictness of speech, composed of different faculties at all, but only as acting in different modes, and to a greater or less extent through different organs. The analysis here suggested is not even a description of the constituent parts of the mind. It is not asserted, nor even assumed, that the mind has different parts. It is simply an analysis for the convenience of classifying the various mental phenomena. The same mind hears and sees ; but hearing and seeing are not the same. So the same mind reasons, imagines, remembers ; but reasoning, imagining, and remembering are not the same. Though for convenience I use the term faculty in this classification, the classification is simply suggested for the better and more orderly arrangement, and more satisfactory study of mental activities as actually seen in real life and living characters.

2. It is not, therefore, necessary for us to consider whether the powers or faculties here mentioned are original and simple powers of the mind or not. They are not suggested as original and simple powers of the mind. In some instances they clearly are not. Mr. Bain has shown, I think, very

clearly, that combativeness and destructiveness may be traced
to a love of power; that they are mainly, if not wholly, manifestations of a love of power. " The feeling of power essentially implies comparison, and no comparison is so effective
and so startling as that between victor and vanquished. The
chuckle and glee of satisfaction at discomfiting an opponent,
no matter by what weapons, are understood wherever the
human race has spread, and are not wanting to the superior
animals." This is true. Nevertheless, the manifestations of
the love of power are so various, that for purposes of classification it is convenient to put by themselves those which are
exhibited in combat and destruction. Something of the
same fundamental motives may underlie the constructive
work of Stevenson and the destructive work of Von Moltke,
but in the study of human nature these different manifestations need to be classified under different titles. So
again, it may be true, that acquisitiveness is not an original
instinct, but is simply the rational endeavor of man to obtain
the advantages supposed to be conferred by wealth, and to
avoid the evils produced by poverty; though this theory
hardly accounts for the blindness of covetousness and the
self-imposed wretchedness of the miser. But whether it is
true or not, it aids in the study of life to recognize acquisitiveness as though it was an original and simple instinct, and
to place under it certain common phenomena of modern
commercial life.

3. For the same reason this classification is not, and no
such classification can be, perfect. It is like an index to a
book; there must be some cross references. It is like a set
of pigeon-holes or envelopes in which the student places his
memoranda and his scraps; some of them he is puzzled where
to put, for they belong in two or three separate compartments, and might go with about equal propriety in either one.
The student, therefore, must not take this classification as
though it were a topographical map of the human mind; a

picture of what the mind is, though possibly to be improved
and corrected by future explorations and surveys. He must
take it as a suggested *Index Rerum*, for the better arrangement
of mental and moral phenomena. If he observes mental and
moral phenomena, which he cannot find a place for in the
tabular view here suggested, he must enlarge it ; or, if he
finds it easier to arrange phenomena here divided into two or
three compartments under one, he is at liberty to omit whatever
seem to him superfluous titles. The main thing for every
reader of this little treatise is to study human nature—in life,
in fiction, in history—for himself, and use this analysis just in
so far as it aids him in that independent and original study,
and no further.

4. It is further to be remembered that even if the mental
and moral powers be regarded as real and separable powers
of the mind, they certainly do not act independently of each
other. We sometimes hear it said of a man, in criticism of
him, that he acted from mixed motives. Every man always
acts from mixed motives. His clashing desires act upon each
other, and his action is the result not of any one impulse, but
of several impulses of unequal force combining together.
Man may be compared to a croquet ball upon the lawn ; the
principal motive to the mallet which gives him a first direction ; but the unevenness of the ground and the other balls
give new and different directions to his activity, and the final
direction which he takes is the sum of all their influences.
Only the more confirmed and inveterate miser acts under the
impulse of acquisitiveness alone. In nearly all men it is
variously modified by self-esteem, approbativeness, conscientiousness, combativeness and destructiveness, benevolence ;
and the conduct of life is in no two men exactly the same,
because in no two men is the sum of their various impulses
the same. In unriddling man the student must take account
of all these various and often antagonistic forces within him.
Thus, for example, when Adam Bede saw Hetty and Captain

2*

Dormithorne kiss and part in the woods, he is described as being in a tumult of contending emotions. If we may turn the drama into cold analytical psychology, we might say that his amativeness or love for Hetty, and his self-esteem or wounded self-love, both of which were strong passions in him, impelled him to punish Hetty's unconscious enemy and his own, while reverence for one socially so much his superior held him in check, and conscience bade him rebuke but not revenge. When at last he gave way, and struck the blow which stretched Captain Dormithorne senseless, it was because for the moment amativeness and self-esteem proved too strong for reverence and conscience; when he stopped to lift up his prostrate foe, restore him to consciousness, and bring him to his home, it was because conscience, reverence, and benevolence—the latter aroused to pity by the helplessness of his enemy—re-asserted their sovereignty once more. Thus no action in life can be attributed to any one faculty. Nearly every action is the result of composite forces.

5. Especially is it important to bear in mind that the lower faculties are affected and often revolutionized in their activities by the higher faculties. No faculty is sinful, and no faculty is free from the possibility of sin; it is the office of religion to make the spiritual dominate the animal and the social nature; such domination changes radically every activity. Thus the animal appetites, if left unregulated, lead to the grossest gluttony; to excesses so bestial that we shudder at the mere recital of them. But those same appetites, restrained by conscience and guided by reason, become the instruments for building up the body in physical health and strength, and making all its organs fit instruments for the mind and soul. The sexual instinct left to itself runs riot in all horrible forms of sensuality and lust. But purified by faith, regulated by conscience and reason, and mated to love, it becomes the most sacred of all earthly ties, and the foun-

dation of the most sacred and essential of all earthly institutions—the family. Whether acquisitiveness becomes an incentive to plundering greed, or productive industry; whether combativeness and destructiveness become incentives to pillage and war, or simply the supports to a great Protestant Reformation or a great war of Emancipation; whether caution makes its possessor a coward and an apostate, or the wise and courageous defender of sacred interests intrusted to him; whether his self-esteem makes him a haughty Gregory the Great, or an unbendable William the Silent, depends upon the presence or absence, the power or weakness of the spiritual faculties, and the consequent influence they exert in transforming the lower nature, and giving its powers a new activity and crowning them with a new life.

With these preliminary explanations I proceed to our analysis.

The most natural division of the powers of the soul is into two great classes: the Motive Powers and the Acquisitive Powers. By the Motive Powers, I mean those which supply motive, force, impulse, power; by the Acquisitive Powers, those which furnish information, knowledge, truth. The Motive Powers again are divided into the Animal Impulses, which are necessary to the support and protection of life; the Social and Industrial Impulses, which make man a social being and underlie his social existence; and the Spiritual Impulses, which are peculiar to him, and distinguish him from the mere animal creation. The Acquisitive Powers again are divided into the Sensuous, the Supersensuous, and the Reflective. This classification, with the suggested faculties under each division, will be found at the end of the book in a tabular form.

CHAPTER VI.

THE ANIMAL IMPULSES.

1. THERE are certain motive powers which are essential to the support of animal existence. These are the appetites necessary to the support of the individual, and the sexual passion necessary to the support of the race. There still lingers in the Church and in religious teachers a remnant of the old Gnostic philosophy which made all sin to consist in the body, and therefore treated all fleshly appetites and desires as sinful. Men still regard appetite and the sexual desire as sinful, because they lead to so much and so palpable evil, and it must be conceded that there are phrases in the New Testament, especially in Paul's Epistles, which, if taken out of their due order and connection, give some color to this view. But the teaching of the New Testament, including that of Paul, if taken in its entirety, gives no warrant to this false philosophy of human life. On the contrary, Paul explicitly warns the Colossians not to be subject to the rule of this ascetic philosophy of life, " Touch not, taste not, handle not ; " and to the Philippians he declares that he knows how to abound as well as how to suffer want. What the Bible condemns is the supremacy of these animal appetites and desires over the intellectual and spiritual nature. They are the lowest of all the impulses, and should be subordinate. When they demand control they are in revolt; when they obtain control the soul is in anarchy. Then the mob has mastery of the palace, and destruction is inevitable. The appetite has for its principal function to induce the individual to take such food and fluid as is necessary to supply the waste of the body and keep it in a good physical condition

Connected with it is a palate which accepts some articles of diet and rejects others. Both the palate and the appetite may become diseased; their action is rarely absolutely healthy, and never infallible; but a desire for a particular article of food is generally a sign—though often a misleading one—that the body needs that particular article, or at least the material which that article supplies. In the case of two boys, brothers, one of whom is very fond of sweets, the other of acids, the desire in each case is an indication of the needs of the two organizations. So a craving for meat in one, and a distaste for it in another, is an indication that the one requires and the other does not require it. If the body were perfectly healthy, and in a perfectly natural environment, the appetite would be a reasonably, possibly an entirely, safe guide. This is not, however, the case. Diseased appetites, unnatural and unhealthy desires, have been handed down from generation to generation. Civilization has brought many influences to bear upon man which produce unnatural desires. A fever produces an intense craving for water, which is due, not to a real want of more liquid in the system, but to an unusual heat which craves cooling. So overwork and overexcitement produce a demand for stimulants; bad air and bad food a demand for too much nutriment, or perhaps a distaste for all. The ill-educated palate requires sweets or spices. The dyspeptic's hunger is no indication of a need of food, and his sense of overfullness is no indication that he has nutriment enough. In a word, the instincts are very far from being a safe and trustworthy guide to be undeviatingly followed. They are symptoms whose real significance the reason must consider and interpret before they can be followed with safety.

2. In a similar manner the sexual passion is essential to the perfection of the race. It repeats and emphasizes the divine command given by God to our first parents: "Be fruitful, and multiply, and replenish the earth." Without it

the family would be impossible. Our children ought to be
early taught by their parents its sacred significance and its
value. They ought not to be left to learn about it from often
immoral and always ignorant companions. They ought not
to be punished for falling into a habit of self-indulgence
against which they have never been warned. This strange,
mysterious desire, which always accompanies health and
vigor, and which prompts both to the purest love and the
most bestial excesses, cannot be eradicated, for God has
planted it in man ; it should be early directed by the child's
natural guardian and teacher. As with the individual, so
with society ; the social evils which grow out of the sexual
appetite are various and deadly. They are often fostered
directly by unscrupulous men for purposes of gain, from motives of avarice. They can be checked somewhat by law
but so long as appetite exists, so long the sins of appetite
will continue to poison society ; and the only real and radical
remedy is that education and that spiritual development
which brings the appetites themselves into subordination to
the law of God as revealed to and written in the higher
nature. Law can protect society from these evils in some
measure ; but no law can eradicate them. Nothing can do
that but the subjection of the appetite and the supremacy of
the spirit.

3. Next to the appetites and passions come the impulses
of combativeness and destructiveness. The former were
necessary to the support of life ; these are necessary to its
protection. Man is surrounded by enemies ; enemies to his
existence and to his progress ; enemies to his physical and to
his spiritual well-being. He needs, therefore, to be endowed
with certain powers of combativeness and destructiveness ;
powers which enable him to stand up strongly, contend
bravely, and destroy utterly. The exercise of these faculties
of combativeness and destructiveness is commanded also to
our first parents in the Garden in the law, " Subdue it [the

earth] and have dominion." This man could not do unless
he were fitted to be a combatant; without both the powers
and the instinct of combat, he could not conquer nature,
subdue the wilderness, battle with the wild beasts, and so
tame the world to be his dwelling-place. For this he must
have the moral as well as the material force; the impulse as
well as the nerve and the muscle. The possession of these
qualities give force, energy, courage, pluck, push. They are
seen in every pioneer, in every great captain, in every man
of large success. Without this power Luther could never
have burned the Pope's bull in the court-yard at Wittenberg;
nor could Paul have faced the mob from the stairs of the
Tower of Antonia; nor could Christ have driven the traders
from the temple courts. It is this which made him the Lion of
the tribe of Judah. A single individual may be an estima-
ble member of society without it; for others about him,
stronger and more courageous than himself, will do his bat-
tling for him, and he will compensate by other and gentler
services. But the human race could not survive its loss. It
would be overborne and perish from its own weakness and
imbecility. This gives power of punishment to all govern-
ments. It is at the root of every form of wrath and indignation. It enables the parent to punish his child; the government to punish crime. It breaks out in lynch law against the
desperado. It may become the instrument of any other
faculty. Serving acquisitiveness, it becomes predatory, and
makes its possessor a robber and a plunderer; serving conscience, it becomes an honorable courage, and makes its
possessor a guardian of the interests of his home or his
state from the robber or the anarchist. The lack of it
begets irresolution, effeminacy, weakness, cowardice; its
excess, or ill-direction, or ill-control, begets quarrelsome-
ness, a disputatious spirit, the gladiator, whether with
muscle or brain, cruelty, rapine, murder. It is indispensable to the existence of mankind, but it is also one of the

prolific sources of all that is inhuman in history and in life.*

4. Akin in its object, but contrasted by its nature with combativeness and destructiveness, is cautiousness. The one protects by fight, the other by flight. The one is the lion in man, the other is the hare. The commingling of the two constitutes true courage; for there is no true courage without a perception of danger and a desire to avoid it. Caution is one of the restraining impulses, holding men back from too sudden, too aggressive, and too heedless action. It compels them to pause, to reflect, to consider. It is a rein; combativeness and destructiveness are spurs. It is strongest in women; is seen in its worst aspects in effeminate men. It is the cause of all cowardice; leads to concealments; is manifested in ordinary social life in the sensitive disposition of the timid; often underlies a vacillating disposition; is the most common cause of deception and falsehood; and should be counteracted always by hope, courage, conscience, love; almost never by severe punishment. It is invaluable as a restraining and counteracting faculty; when it becomes the dominant motive, it is fatal to forcefulness and efficiency of character.

5. Among the impulses whose object is a preservation of existence must also be put the love of offspring. So much has been said and written about parental love, about mothers' love especially, that it may seem to the reader doubtful whether this impulse belongs here among the lower animal impulses. But a moment's reflection will convince him that the love of offspring is in its lowest forms a purely animal instinct; seen in the cat's care for her kitten, the hen's for her chickens, the cow's for her calf in every farm-yard; seen also, alas! as a mere blind semi-sensual instinct, in many a

---

* The phrenologist generally distinguishes between combativeness and destructiveness. But they are so nearly akin, that I think any discrimination between them is rather confusing than helpful in analysis.

home, where the father or mother cannot bear to inflict pain, or thwart a desire, or permit a disappointment, or allow a burden, and so the child grows up, coddled and tended, to be weak and wayward and willful, and often worse. This parental instinct, guided and inspired by the higher nature, is the child's guardian from present evil, and guide into future manhood; but unguided and uninspired, it protects only from pain, which is God's method of discipline, and seeking only happiness, guides often into destruction and misery. It is, too, quite evident that it is necessary for the protection of existence ; for the infant, whether of man or animal, is rarely able at first to protect himself; the higher his rank in the scale of being the greater the necessity for protection ; and if there were no parental instinct, if there was nothing but a general and distributed sentiment of pity, he would certainly suffer greatly, and would generally die for want of the power in himself of self-protection. The parental instinct endows him with all the faculties and powers of his parent, especially with those of his mother—for in both brutes and men this instinct is almost invariably the strongest in the female—until his own powers have attained sufficient growth to make him able to protect himself.

CHAPTER VII.

THE SOCIAL AND INDUSTRIAL IMPULSES.

1. AT the foundation of the social organization in all its various manifestations is the social instinct. The phrenologists call it adhesiveness. Some animals are gregarious, others are solitary. Man is gregarious. The recluse is an exception. There is but one Thoreau. Solitary imprisonment is the most dreaded of all penalties. Men are impelled to associate together in political, industrial, and social enterprises. Their intercourse in these associations is regulated in a large measure by the spiritual impulses, of which we shall have something to say in the next chapter. But the association itself is a necessity of human nature irrespective of the ulterior advantages to be gained from it.

Bain attempts, but not very successfully, to account for the social instinct by the fact that it is a means to an end. We associate, according to him, to gratify our benevolent impulses; to get aid from others in our life and its undertakings; to gratify our love of power or of applause, and the like. These most certainly intensify the social instinct; but the social instinct exists independent of them. A man may be very social and yet supremely selfish; he may dread isolation and yet be cynical. Sociality is a primary fact of human nature. There is a molecular attraction which draws men together. Humanity instinctively coalesces as drops of water in a stream.

2. Doubtless one of the chief promoters and regulators of this social instinct is approbativeness, or the love of praise. Mr. Darwin regards it, I think wholly without good ground, as constituting the basis of conscience. But unquestionably

it re-enforces conscience in some cases, and serves as a very poor substitute for it in others. It promotes, if it does not produce, what we commonly call good nature. This is very different from benevolence. The good-natured man does not live to produce the greatest amount of good to the greatest number, nor even to his immediate neighbor. He tries to please him not to build him up, and is often as ready to do him a real injury as a real benefit, if it will win the reward of approbation at the time. He that is governed by approbativeness is a prey to many and sometimes seemingly diverse faults. He is easily discouraged, and not easily satisfied. The more he is praised the more praise he demands. He is guided and often governed by the opinions of others. He is rarely strong unless his approbativeness is more than balanced by some other motive, self-esteem, for example, or conscientiousness. This is the secret source of the human passion for " glory ; " for it, as expressed by a medal, the scholar toils and the soldier fights. It is universal ; perhaps of all the motive powers the most various in its activity and the most pervasive though not the most powerful in its effects.

> " The love of praise, howe'er concealed by art,
> Reigns more or less and glows in every heart ;
> The proud to gain it, toils on toils endure,
> The modest shun it, but to make it sure."

When it is the foundation of character it lures on to inevitable moral ruin. The man who makes approbativeness a substitute for conscience has no other standard of right and wrong than the opinion of his own public, which is a far less trustworthy standard even than public opinion. He becomes a chameleon, changing his character and his opinions to suit the company he is in, and doing it instinctively, unconsciously, and in a sense honestly. When he loses the approbation of his fellow-men he loses the last restraint and incentive. The history of Aaron Burr is the history of a man of brilliant parts, wrecked by the absence of a conscience and the sub stitution as a moral mentor of approbativeness. On the other hand, he who lacks approbativeness lacks tact, sympathy, quick fellowship, readiness to appreciate, or willingness to weigh, the feelings and opinions of others. He does not assimilate with others, for he is indifferent to their sentiments. He is disregardful not always of their real interests, but of their wishes and feelings. He does not consider the effect of his example on others, and so allows his good to be evil spoken of. His self-esteem becomes an intolerable self conceit, and his conscience a tyrant over others. He becomes pert, angular, rude, boorish.

3. Self-esteem is sometimes popularly confounded with approbativeness; it is in actual experience more commonly the antidote thereto. Approbativeness leads us to desire the approbation of others; self-esteem leads us to desire our own. Approbativeness asks what will others think of us; self-esteem, what shall we think of ourselves. Self-esteem tends to give its possessor independence of thought, individuality of action; to make him forceful and vigorous. It is to be found in nearly all born leaders, whether of thought or of action. Its normal and natural exercise produces self-reliance and enforces courage. If it is not excessive, it is a consciousness of power and adds to real strength of character. If it is excessive, it is an imaginary consciousness of power which has no real existence, and is a fatal weakness. The divine law of self-esteem is expressed by the apostle Paul in the direction to every man, " not to think of himself more highly than he ought to think; but to think soberly." A wise, right, and true estimate of one's own powers is necessary to their highest and best use. The general who overestimates his forces leads them to defeat; he who underestimates them does not lead them at all. Self-esteem in excess leads to pride, censoriousness, arrogance ; it makes its possessor impervious to criticism, and even to

the lessons of bitter experience. Whatever his failures, he always attributes them to other causes than his own mistakes. He is always too wise to learn. It is the exact opposite of that poverty of spirit which Christ said was the first condition of entering the kingdom of heaven. " Seest thou a man wise in his own conceit; there is more hope of a fool than of him." On the other hand, the lack of a due and proper self-esteem is almost invariably accompanied with an excess of approbativeness. The result is a weak and vacillating character. A man who has no confidence in the wisdom of his judgments or in his power to execute them is timid, irresolute, uncertain, dependent upon others, a follower of stronger natures, never a leader. Approbativeness is the most common weakness of women, self-esteem of men. The one neutralizes the other. Both cannot well be in excess. He who has an excessive confidence in his own powers will be indifferent to the commendation and criticism of his fellow-men; he who has a supreme regard to the opinions of his fellow-men will be distrustful of his own. Neither approbativeness nor self-esteem are necessarily a fault. An illustration of a divine form of approbativeness is afforded by Christ's parting request to his disciples to preserve his memory through all coming ages by a memorial supper. An illustration of a divine form of self-esteem is afforded by his declaration to the Pharisees, "Ye are from beneath, I am from above," and by his declaration to his disciples, "Ye call me Master and Lord : and ye say well; for so I am."

4. The love of acquisition appears to be a primary and instinctive impulse of the human soul. It is, indeed, seen in some of the higher animals, especially in the bee and the ant; but it belongs to the social rather than to the animal nature; it is pre-eminently human. Unquestionably this desire is stimulated by the advantages conferred by acquisition; it is a means to an end ; and from the boy struggling for a toy, to the man struggling for a larger bank account, some

real or imaginary pleasure derived from the possession is generally and perhaps always in view. But there appears to be a pleasure in new possessions apart from the uses and enjoyments which they afford. At all events, whether acquisitiveness be regarded as a primary and fundamental motive power, or only the working out of other motive powers of the mind, it enters so largely into human life that it may well be regarded as a distinct impulse. It is the mainspring of industry; the secret power of material civilization. It is the love of acquisition which has opened a highway for commerce across before untrodden seas, bridged the continents with iron highways, opened the hidden wealth in gold and silver and iron mines, brought lumber from the forests and coal from the hills, plowed the prairies and harvested the plain where the bison once roamed, founded cities on the site of the wigwam. It has been a greater motive power than conscience; it has achieved more for mankind than benevolence. It has sheathed the sword and forged the plowshare; has conquered combativeness; has turned man from a wild beast into a domestic animal, from a destroyer into a producer. But this is all. It is without a moral character or a moral purpose. Guided by reason it avoids criminal dishonesties, because reason sees that wealth acquired by methods which arouse the indignation of mankind is never permanent. But it is the secret of covetousness and avarice. It is the parent of the gambling-house and the liquor-saloon. It is embodied in all speculative operations. Unrestrained by conscience it is dishonest, untempered by benevolence it is cruel. It is a root of all evil. They that are controlled by it, that *will* be rich, fall into a snare. It is at once the most useful and the most despicable of the social motives.

5. Closely allied to it is the constructive instinct; the inclination to build, to put together; prominently manifested in all engineers, engravers, and mechanics. Employed in the intellectual realm its work appears in more subtle forms.

The writer who possesses it constructs his work with skill, though it may be barren of ornament or fancy; the preacher endowed with it makes a good skeleton though he may utterly fail to infuse it with the life of human sympathy and feeling. It is the secret of all modern mechanics, and of very much of modern practical science. It is the opposite of destructiveness; is witnessed in any eminent degree only in the higher stages of civilization; and is, perhaps, the highest of all the social instincts. Its manifestations are witnessed in lower forms in the animal creation, as in the bee and the beaver; but it is essentially a human and social instinct.

6. The tendency to imitate, of which in the animal creation illustrations are seen in the parrot, the mocking-bird, and the ape, is also a human instinct. It is seen in all forms of mimicry; enters largely into every phase of dramatic representation, from that of dramatic oratory, like that of John B. Gough, to that of the stage; underlies much of the power and beauty of art in its less exalted forms; and is the basis of much of our educational system. It is this inclination on the part of humanity, especially of the young, to imitate the actions of others, which gives such power both for good and for evil to example. It is this which makes leadership possible; for there can be no leadership without an inclination to follow the leader. It induces men to take their opinions from others; to copy the actions of others; to model their characters after others. It is strongest in the young, or in the crude and uneducated. John Chinaman given a plate as a model, in which there happens to be a crack, makes each of the dozen with a corresponding crack. The boy smokes, not because he likes the cigar, which sickens him on a first and even a second or third trial, but because he sees his elders smoke. Approbativeness and imitativeness work together. They are harnessed as in a span. Without the instinct of imitation, society would tend to lapse into a mere congeries of individuals; it would learn from the experience

of the past very slowly, if at all.   On the other hand, excessive imitativeness destroys all individuality and independence of character, and reduces the man to an automaton, who moves only in drill, and does nothing except in blind imitation of a supposed superior.

7. To these social instincts should probably be added also the instinct of local attachment.   It is certain that some persons become very strongly attached to places; others have no such attachment.   It is said that the cat is attached to the house, the dog to the master.   The one pines for the house, the other for the man.   A like difference is seen in men.   Generally women have stronger local attachments than men.   A rude violation of this instinct is one of the chief causes of home-sickness.   A principal value of it is a certain kind of local stability.   Without it all men would be, as are the Bedouin Arabs, nomadic.

CHAPTER VIII.

THE SPIRITUAL IMPULSES.

MAN is distinguished from the rest of the animal creation by his moral and spiritual nature. The distinction recognized in the earlier books of mental science between reason and instinct is now largely abandoned. There are instinctive and almost automatic actions, as there are intelligent and thoughtful ones; but the distinction between the brute and the man is not in the possession of mere instinct by the one and of reason by the other. Man sometimes acts from instinct; he does so whensoever he follows blindly one of the impulses which we have described above, without stopping to submit the proposed action to the questions and directions of his reasoning powers. On the other hand, there are abundant evidences of the possession and use of reasoning power by the brute creation, though in very crude forms and within very narrow limitations. The dog, the horse, the elephant, consider, reflect, reason. They exercise the faculty of causality and comparison, of which we shall have something to say in a subsequent chapter. But there is no indication whatever of the possession or exercise by them of any moral discrimination, or of any spiritual power. It is the moral and spiritual powers of which we are now about to speak which distinguish man from the brute. Brutes reason as truly as men; but every man is a law unto himself, while the brutes are subject to law only as they are brought under it by a superior force. They do not rule themselves by recognized laws of right and wrong. The only law recognized is the law of the strongest. All men worship; the exceptions, if there are any, are so few as to be insignificant. Every nation

3

has its religion or its superstition, its god or its demon, its temple or its fetish. There is no indication of any thing analogous to worship among the brutes; they have houses, but no temple; social organization, but no revealed law; domestic instincts, but no spirit of universal benevolence.

1. Conscience is the factor which recognizes the inherent and essential distinction between right and wrong, and which impels to the right and dissuades from the wrong. It does not come within the province of this book to discuss either the basis of ethics or its laws; to consider either why some things are wrong and others are right, nor to point out what is wrong and what is right. That belongs to moral science, not to mental science. It must suffice here to say that the distinction between right and wrong is recognized in all peoples, and is one of the first objects of perception in childhood. Standards differ in different races and in different ages. The power of moral discrimination is subject to education both for good and for evil. But the sense of ought is as universal as the sense of beauty. That there is a right and a wrong is as evident to every mind as that there is a wise and a foolish, a beautiful and an ugly, a pleasant and a disagreeable. There are things pleasant and things repulsive to the moral sense, as there are things pleasant and things repulsive to the eye and to the palate; there is a sense of right and wrong as there is a sense of beauty and a sense of taste. It has been a matter of great debate among philosophers what is the ground of right and wrong. We cannot here enter into this debate. I shall assume, what is by no means universally conceded, that it is a primary fact in life; that the right is right and the wrong is wrong, irrespective both of commands and consequences; that the right is right not because God commanded, but God commands it because it is right; and it is right not because it produces happiness, but it produces happiness because it is right. That it would still be right though it produced misery instead of happiness,

and was forbidden, not commanded; that it is as truly the law of God's nature as of man's nature; and that if we could conceive his commanding any of his children to do what is not right it would change, not the character of the action, but his own character. I assume, too, that as right and wrong are primary facts of human life, so the faculty which recognizes that fact, and which impels men to do right and to eschew wrong, is a primary faculty. Men are to be guided by their judgment in determining what is right and what is wrong; but the judgment does not determine that there is a right and a wrong. Their sense of right and wrong is clarified or obscured, their impulse to the right and away from the wrong is strengthened or weakened, by other faculties; but it is not dependent upon them. Approbativeness may lead them to do what other people think to be right; but desire for the approbation of others is not conscience, nor the ground nor basis of conscience.* Self-esteem may strengthen their purpose to do right, and so win their own approval; but self-esteem is not conscience, and self-esteem and conscience may come into direct conflict. Benevolence may add its persuasions to the impulse of conscience, and the man may be impelled to do right because doing right will also do good to others. But this is not the ground of his conviction that there is a right; and the right may even seem to be fraught with irreparable injury and no compensatory good to others, and so benevolence and conscience come in conflict. The recognition of right and wrong and the impulse to right and away from wrong is original, primary, causeless, one of the simple and indivisible powers of the soul of man. I cannot better state this truth—I have no space here to argue it—than in the words of Professor Huxley, who will certainly not be accused of any undue orthodox proclivities:

---

* As Darwin makes it. See his " Emotions in Animals and Man."

" Justice is founded on the love of one's neighbor; and goodness is a kind of beauty. The moral law, like the laws of physical nature, rests in the long run upon instinctive intuitions, and is neither more nor less innate and necessary than they are. Some people cannot by any means be got to understand the first book of Euclid; but the truths of mathematics are no less necessary and binding on the great mass of mankind. Some there are who cannot feel the difference between the Sonata Apassionata and Cherry Ripe; or between a grave-stone cutter's cherub and the Apollo Belvedere; but the canons of art are none the less acknowledged. While some there may be who, devoid of sympathy, are incapable of a sense of duty ; but neither does their existence affect the foundations of morality. Such pathological deviations from true manhood are merely the halt, the lame, the blind of the world of consciousness; and the anatomist of the mind leaves them aside, as the anatomist of the body would ignore abnormal specimens." *

It is hardly necessary to point out the function or dwell upon the necessity of conscience in human life. It is fundamental to all that is human in life. Without it men would be brutes; society would be wholly predatory ; the only law recognized would be the law of the strongest; the only restraint on cupidity would be self-interest. There could be neither justice nor freedom. Trade would be a perpetual attempt at the spoliation of one's neighbor. Law could be enforced only by fear, and government would be of necessity a despotism. The higher faculties uninfluenced by conscience would rapidly degenerate. Reverence would no longer be paid to the good and the true, but only to the strong and the terrible ; religion would become a superstiticn ; God a demon ruling by fear, not by law ; punishment a torment inflicted by hate and wrath, not a penalty sanc-

---

* " English Men of Letters : Hume," by Prof. Huxley. Harper & Brothers, p. 206.

tioned by conscience for disregard of its just and necessary laws; and benevolence itself, unregulated by a sense of right and wrong, would become a mere sentiment, following with its tears the robber as readily as the Messiah to his crucifixion, and strewing its flowers as lavishly on the grave of the felon as on that of the martyr. In history have been seen all these exhibitions, not of the absolute elimination of conscience from human life, for conscience has never been wholly wanting in the most degraded epoch of the most degraded nation, but of its obscuration and its effeminacy.

It is, perhaps, more needful to remark that the evils of a character whose conscience is not controlled by other and still higher faculties are quite as great. Conscience combined with self-esteem, uninstructed by faith and unrestrained by benevolence, is the most remorseless and cruel of impulses. The ingenious persecutions which it invented and inflicted in the Middle Ages on the Protestants of Italy, Spain, France, and the Netherlands, far exceeded those which mere brutal combativeness and destructiveness inflicted on the Jews in the reign of Nero. It is not enough that a man be conscientious. He may be conscientious and self-conceited; in that case he will be exacting and despotic, making his own conscience a law for all his neighbors. He may be conscientious and approbative; in that case he will be weak, afraid, and always tormented lest he has not done what to his neighbors will seem to be right. He may be conscientious without faith; in that case he will be constantly led into false judgments by a tendency to measure the moral quality of every act by its immediate effect. He may have a merely retroactive conscience; in that case he will fail to look forward and prepare for what he is to do by measuring proposed action by the standard of right and wrong; he will be habitually looking back and tormenting himself, and perhaps others as well, by perpetually trying himself for actions past and beyond recall. He may use his conscience not as the restraining motive of his life, but as the impelling motive, not as the governor, but as the steam; in that case he will have nothing of the joy of the perfect love which casteth out fear, but will always act under the spur of necessity, never in the freedom of those who through faith and love have entered into the liberty of the sons of God. It is not enough to have a conscience, and a masterful conscience; it must be a good conscience; a conscience that forecasts; that acts in restraining rather than in impelling; that is instructed by faith, not by sight; that is united to benevolence rather than to approbativeness and self-esteem.

2. Reverence. There is in man an instinct inclined to look up, to admire, to reverence. Something akin to it, something certainly illustrative of it, is seen in the apparent mental attraction of the best and most intelligent dogs toward their masters. But in the brute it is apparently dependent largely on physical services rendered and on fear. It is seen in man in various forms in the social organism, and is in one sense a social instinct, as also is conscience. But in its higher manifestations it is essentially both human and spiritual. It is the basis of all wonder, admiration, awe, reverence.* It is the foundation of that awe which we feel in the presence of the great and sublime in nature: the vast wilderness, the towering mountain, the starry heavens. Fear sometimes, but by no means always, enters into its existence; the two emotions are indeed often in absolute contrast, so that one hardly knows whether to fear or to rejoice. It enters into our experience of admiration of human handiwork, in art, mechanics, architecture. It is the basis of social distinctions, especially as they are seen in countries where hereditary classes exist, and the lower class is habituated to looking up to the class above it, where looking *up* is as easy and as natural

---

* The phrenologists recognize two faculties, one of reverence, the other of marvelousness. This seems to me a needless and doubtful distinction. Essentially and at root they are the same.

as for an American to look *off.* It is the instinct of the child toward the parent, making it easy for the one to obey and the other to enforce the command, "*Honor* thy father and thy mother," a command often read as though it were, what it is not, *Obey* thy father and mother. It is seen in every form and phase of worship. It impels men every-where to a belief in some superior Being, known or unknown, imagined or unimagined, but deserving and demanding and receiving reverence. It exhibits itself alike in the devotee bowing before his hideous image in his magnificent Hindu temple, in the Friend lifting up his heart in silent adoration to the invisible Spirit, and in the spirit of wonder and of awe with which the seemingly undevout scientist approaches the confines of the visible world and looks off and seeks to fathom the beyond, and returns shaking his head in intellectual despair; saying it is the Unknown and the Unknowable. It is ignored by atheists, and therefore atheism has never made many converts, and never can. It is an essential and indestructible principle of human nature.

But this faculty is no more free from dangerous propensities than the lower social and industrial impulses. It is as dangerous when ill-educated and misdirected as acquisitiveness or self-esteem. If for lack of it men grow skeptical, infidel, atheistic Godward and cynical manward, inclined to take low views of man and none at all of God, its excess leads to idolatry and every form of superstition, to worshiping the unworshipful and reverencing that which is not venerable. Mated to conscience and self-esteem it produces bigotry; uninstructed by faith it begets idolatry or the worship of the visible, and therefore the unreal; combined with fear it begets superstition, the worship not of what is to be venerated, but of what is to be dreaded.

3. Benevolence. By benevolence is meant the impulse which leads its possessor to wish well to all other beings. It is an innate, not an acquired, quality of the mind. It exists

in all men, though in many buried and almost destroyed by the pre-eminence of other faculties. It has its weak side and its defects. Uninstructed by faith it desires merely the happiness, not the welfare of men, and sacrifices without hesitation their real and permanent good for their apparent and present pleasure. Mated to approbativeness it becomes a mere instinct or impulse of good nature. Unregulated by conscience it is indiscriminating, a mawkish and morbid sentimentality. But it is of all the impulses the one whose vices are the least dangerous, and whose virtues are the most beneficent to mankind. Coupled with veneration, and looking up to a superior, especially to God, it redeems worship from fear, and makes it ennobling, elevating, purifying. Looking upon suffering, it is pity; looking upon sin, it is mercy; looking upon the well-being of the whole community needing protection from sin, it is justice. As an emotion, it is sympathy; it weeps with those that weep, it rejoices with those that rejoice. As a principle of action, it seeks the greatest good of the greatest number. It combines with the social instinct to make the family of man more than the nest of birds, society more than an ant-hill or a bee-hive. It is the secret of patriotism in the State, and is that love which is the bond of perfectness in the Church. It is the queen of the soul, and he only is truly healthy whose whole nature is obedient to love—whose acquisitiveness gathers for it, whose combativeness and destructiveness battle for it, whose caution fears for its wounding, whose conscience is made tender and sympathetic by it, and whose reverence is made fearless and filial and joyous by it. It is incapable of analysis; and no description which has ever been penned of it can compare for intelligent comprehensiveness and spiritual beauty to Paul's psalm of praise to it in the thirteenth chapter of 1 Corinthians: "Love suffereth long, and is kind; love envieth not; love vaunteth not itself, is not puffed up, doth not behave itself unseemly, seeketh not its own is not provoked, taketh not account of evil; rejoiceth not in unrighteousness but rejoiceth with the truth ; beareth all things, believeth all things, hopeth all things, endureth all things."

4. Phrenologists add to the impulses thus far given several others, the two most important being hope and firmness. These seem to me to be rather qualities than faculties; not so much impulses to do particular things as a spirit or habit which modifies the method of doing all things. Hope gives vigor of plan, firmness gives continuity of purpose ; the one makes radiant, the other makes strong. Neither are simple or primary passions. Both are capable of analysis. Hope is desire and expectation commingled. What a man desires but does not expect, he does not hope for; what he expects but does not desire, he does not hope for; what he both desires and expects, he hopes for. His expectation may be rational or irrational ; his hope is accordingly well or ill grounded. Firmness is vigor and persistency combined. It is persistence of force. Vigor without persistence produces dash and impetuosity; persistence without vigor produces inertia. Health of body has much to do with both qualities. Certain diseases invariably produce despondency. Certain physical weaknesses invariably produce more or less vacillation of purpose. Consciousness of power has much to do with both qualities. The man who is conscious of his own resources will be hopeful of results when all around him are in despair; he will be persistent in his purpose when all about him are discouraged and ready to retreat. He may be hopeful and firm in certain directions, despondent and vacillating in others; firm where conscience is concerned, and weak where self-interest is concerned; firm in protecting the interests of others, and weak inprotecting his own, or *vice versa*. Whether, however, hope and firmness be regarded as impulses, or as temperamental conditions affecting all the impulses, they must certainly be taken into account in any study of human nature.

3*

CHAPTER IX.

THE ACQUISITIVE POWERS. .

I. *The Senses and the Supersensuous.*

WHAT are very generally called the Intellectual Powers
I call the Acquisitive Powers, because among them is a
power which is more and higher than the Intellect; which
at all events it is wise for us to distinguish from the Intellectual, as that phrase is ordinarily used.

1. *The Senses.* There are five of these: sight, hearing,
smell, taste, and touch. Some philosophers have, indeed,
undertaken to show that all the senses are only refined and
subtle forms of the sense of touch: invisible ether striking
the eye, waves of air striking the tympanum of the ear, fra-
grant emissions touching the delicate nerve of the nose, and
subtle qualities in the food coming in contact with the nerves
of the palate. But for all practical purposes the old distinction between the five senses may be accepted as satisfactory.
Nor do these five senses require any explanation. We are
all measurably familiar with them and their exercise.

2. *The Sensuous Faculties.* With these senses are connected sensuous faculties which receive and appropriate and
appreciate the various facts brought to the knowledge of
the mind by the senses. We not only perceive the outer
world, we appreciate qualities in it which the senses themselves know nothing of. Thus a man on horseback emerges
from a forest upon the top of a high hill, and both look off
upon the prospect of the valley below. The same picture is
impressed on the retina of both beast and man, but the same
emotion and intellectual life is not awakened in both. There
are in the rider faculties which perceive in the beauty of the

view that which is unperceived and unperceivable by the horse. Men are said to have or to lack an ear for music; the ear itself is struck by the same sound waves, and the same impression is made on the auditory nerve of both, but the one receives what the other does not and cannot receive. So two men may look at the same picture, and see it with equal distinctness, but the one who has an eye for art will perceive in it much which the other, who has no eye for art, fails to perceive. I shall not enter into any elaborate analysis of the sensuous faculties, nor into the discussion which has been waged concerning them. It is enough for my purpose in this treatise to point out the more evident ones. These are time, space, weight, color, and tune. The powers which recognize these qualities or realities are peculiarly human. Their recognition is very dim, if it exists at all, in the brute creation; it is very clear and absolutely necessary in man. We are compelled to think of all events as occurring in time, and having a certain time relation to each other, as being past, present, or future, as occurring before or after. We are compelled, too, to think of them as occurring in space, and as having certain relations of locality to each other, as being above or below, on this side or that, in this locality or another locality. Whether time and space are intuitive perceptions, or laws of thought binding upon us which compel us to think of things in these relations, or results of observation and experience, it is not necessary for us here to inquire. In point of fact the recognition of both time and space are absolutely universal; the power of recognizing them exists, however its existence may be explained. The same thing must be said of weight, number, color, and time. We have the power of discriminating substances not merely by their form, that is, the space they occupy, but also by their weight or tendency to fall toward the center of the earth. We have a power of recognizing numbers, of perceiving the difference between one and more, of knowing the relations between these various numbers, and of working out of that power of perception the whole science of numbers in all its branches. We have the power of distinguishing colors, and appreciating what we call beauty of color, that is, the combinations of color which tend to produce pleasure through the eye on the mind. We perceive in certain combinations of sounds an effect which we call musical. Moreover, these powers differ very greatly in different men, and seem capable of still further analysis. In tune there are a variety of effects which seem to differ in kind as well as in degree, and the capability to produce or to enjoy these different effects, respectively produce different schools in music, each having its own peculiar and characteristic appreciation. To these should be added the faculty of language, a faculty quite peculiar to man, and differing very widely in different races and in different individuals of the same race.

3. *The Supersensuous Faculty.* We have arrived at a parting of the ways. That there is any supersensuous faculty much of modern philosophy positively denies ; the existence of such a faculty still more of modern philosophy ignores. All forms of modern skepticism have a common philosophical foundation. Their philosophy denies that we can know any thing except that which we learn through the senses directly, or through conclusions deduced from the senses. We know that there is a sun because we see it; we know approximately its weight and its distance from the earth, because by long processes of reasoning we reach conclusions on those subjects from phenomena which we do see. What we do not thus see, or hear, or touch, or taste, or smell, or thus conclude from what we have seen, or heard, or touched, or tasted, or smelled, is said to belong to the unknown and unknowable. This is the basis of modern skepticism. It is the basis, too, of much of modern theology. It is the secret of the " scientific method." By this method we conclude the existence of an invisible God from the phenomena of life ex-

actly as we conclude the existence of an invisible ether from the phenomena of light. But the God thus deduced is like the ether, only an hypothesis. It is quite legitimate to offer a new hypothesis; and the scientist will be as ready to accept one hypothesis as another, provided it accounts for the phenomena. This philosophy, pursued to its legitimate and logical conclusion, issues in the denial that man is a religious being; or possesses a spiritual nature; or is any thing more than a highly organized and developed animal.

Over against this philosophy of human nature I set here the doctrine that man possesses a supersensuous faculty.* By a supersensuous faculty I mean a power to see the invisible and hear the inaudible; a sixth sense; a spiritual perception; a capacity to take direct and immediate cognition of a world lying wholly without the dominion of the senses. A man hopelessly blind might well conclude that there is such a phenomenon as color from the testimony of his friends. A man wholly deaf might well conclude that there was a phenomenon of sound from merely observing its effects on others. But the phenomena of sight are directly and immediately perceived by the eye; they are not ordinarily derived from observation made only by the ear. So I suppose the facts or phenomena of the spiritual life are directly and immediately perceived by the spiritual sense; they are not derived from observation made by the other senses. The spirit has its eye and its ear. This power in art and literature is called the imagination, fancy, ideality; its productions are called creations. In religion it is called faith. There is no better definition of it than that afforded by the author of the Epistle to the Hebrews: " Faith is the substance of things hoped for, the evidence of things not seen."† That is, it is the faculty or power which gives us

---

* Not as though it were in any sense a new doctrine; it is older than Plato, and has never been without its representatives and advocates in philosophical thought.
† Hebrews xi, 1.

knowledge of the invisible realities of life. Religion does not depend upon a "scientific method." God is not an hypothesis. He is known directly and immediately by spiritual contact, spiritual perception.

This power of direct perception of a world beyond the senses is seen in all great inventors. It is the prophetic faculty, and the secret of all progress. Morse sees the telegraph wires carrying his messages over a continent before a pole has been set in the ground. Stephenson sees England covered with a net-work of railway before a rail has been laid. The architect sees his cathedral in the mind before a stone is put to rest in its bed of mortar. In all these and kindred cases the invisible is seen before it becomes visible ; the supersensuous sense perceives it before it can be made apparent to the slower appreciation of the senses. It is the power which underlies all art, literature, oratory. To imitate whether a greensward or a silk dress is not art ; as to declaim whether Marco Bozzaris or Hamlet is not oratory. Art is essentially creative. It brings out of the invisible world invisible realities, and so presents them that the dull eyes and ears of the unspiritual can perceive them. The artist sees his picture before he paints it, and if he be a true artist always sees a nobler picture than he paints. He copies from an invisible canvas. The author sees the truth which he endeavors, always with imperfection, to express; and beneath his dead body of a book there beats a living soul which looks out through its pages as the soul looks out through the eyes into the windows of another soul. This living soul he sees and knows just as clearly before as after he has given it a body. The orator, more dimly but as truly, sees the truth which he endeavors to carry away captive from its shadowy land, and his power over his audience lies in his power to interpret to their senses and through their senses what he had before seen and grasped by his supersensuous faculty, his spiritual perception, his faith-power. These invisible realities thus

seen by the soul's sense are not mere copies of something
which the eye has seen before ; they are not memories, nor
mere new combinations of objects familiar to the senses.
They belong to another world. The artist, the author, the
orator is a true translator into sensuous forms of supersensuous realities, and always views his best work with a sense of
dissatisfaction, knowing that no sensuous forces are adequate
to expound to men who live in the senses what he has seen
and known. He is ready to exclaim with Jesus: "We speak
that we do know, and testify that we have seen; and ye receive not our witness." In still higher forms evidence of
this supersensuous faculty is seen in all our social and domestic life. Our business, our government, our society, our
homes, are all built upon it. Without it they must dissolve
and humanity go back to barbarism and anarchy. All modern commerce is dependent on the reality and the incalcu-
lable value of honor, humanity, integrity; qualities not seen,
not easily demonstrable by a "scientific method," but recognized by all men who possess them and imitated by many
who do not. Justice, truth, honor, fidelity, courage, patriotism, are all intangible, invisible qualities. They are not
seen; they are not deduced from the seen ; they are instantly and immediately recognized as realities by the supersensuous sense. Their value is depreciated or ignored by sen-
sual men. They are qualities unrecognized by the brute.
This faith-power is the recognized life of the home circle,
and of all friendships and fellowships. The love of a mother
for her child is different from the love of a bird for its young,
or a cow for its calf. The love of husband and wife for each
other is more than an animal instinct. The tie which binds
friends together is not sensuous. Identity is not in the feat-
ures. What we love is the inward, the soul, the mental and
moral qualities, the patience, gentleness, forbearance, long-
suffering love, the invisible manhood and womanhood within,
which the eye does not see, which the reason does not dem-

onstrate, which are not hypothetical, which are not ascertained by any "scientific method," but which are instantly and directly and immediately perceived by the power of spiritual perception which resides in every spirit.

But in its highest manifestations this supersensuous faculty is seen in the religious life. It is the power which the Bible calls faith. Faith is not an intellectual activity deducing conclusions from premises; it is not an act of the will or an impulse of the affections, though it inspires both. It is a spiritual perception, "the substance of things hoped for, the evidence of things not seen." By this we perceive the Spirit of God behind all nature and immanent in all nature, as we perceive the spirit of a man behind the body and immanent in the body. Neither are hypotheses to account for phenomena; both are *facts* instantly and immediately perceived. This is the power of which Paul writes when he says, "We look not at the things which are seen, but at the things which are not seen." This is what he generally means by the word *know :* "we *know* that the law is spiritual; " " we *know* that all things work together for good to them that love God;" "I *know* whom I have believed." In these and kindred passages he speaks not of conclusions reached by a "scientific method," but of facts realized by a spiritual experience. It is to the contrast between the sensuous and the supersensuous, between faith and sight, that Christ refers when he promises to his disciples another Comforter whom the world cannot receive, because it seeth him not, neither knoweth (hath experience of) him; "but ye know him, because he *dwelleth with you, and shall be in you.*" It is to this spiritual sense, dormant in even the most unspiritual natures, that Paul refers when in Athens he classes himself sympathetically with his pagan audience, saying, "In him we live, and move, and have our being." This faith-power is the illuminating and transforming power of the soul. It is that whereby God enters it and makes it his own. It is that whereby each faculty is lifted up from a mere earthly and sensuous activity. By faith love is converted from a mere wish for happiness into a wish for true welfare; reverence is changed from image worship to spiritual worship; conscience is able to measure the issues of right and wrong by their intrinsic and spiritual nature, not by the anticipated consequences of action; the parental instinct is lifted above a mere animal propensity, and is made to become a guide to God and a guardian for eternity; and the very appetites and passions are made to minister to the higher, the internal, the spiritual nature.

CHAPTER X.

THE ACQUISITIVE POWERS.

II. *The Reflective Faculties.*

MAN possesses not merely the power to gather both from the outer and visible and from the inner and invisible world, a power both of sensuous and supersensuous observation ; he possesses also a power of classifying and arranging the results of his observation, of observing resemblances and contrasts, and of drawing conclusions from them. This he does by the reflective faculties, or what is in popular language called the reason. Formerly it was supposed that the animals did not possess this power. A more careful and candid observation has brought scientific men to the conclusion that the higher animals—notably the dog, the horse, and the elephant—also reflect, consider, weigh, judge, compare; in a word, reason, though to a very limited extent and within very narrow bounds. The contrast between the animals and man is not that man possesses reflective faculties and the animals do not, but that man possesses apparently an unlimited capacity of developing both these and other faculties, while the limits are very soon reached in the animal; and man possesses the spiritual faculties—the supersensuous faculty of faith and the spiritual impulses of conscience, reverence, and love—in a high degree, while they are either entirely wanting in the brute, or exist only in the most rudimentary forms. For convenience of analysis the reflective faculties may be divided into two, the Logical faculty and the Comparative faculty, or causality and comparison.

1. It does not make much practical difference whether we say that man possesses a faculty by which he perceives the

relation of cause and effect, or that he is under a mental law which compels him to think of all phenomena as in the relation of cause and effect, or that the truth that every effect must have a cause and every cause an effect is intuitively and immediately perceived by him, or that the relation of cause and effect has been perceived by observation and experience through so many generations that he has come to expect an effect from every cause and a cause for every effect as the result of generations of experience. He not only possesses the power, he is laid under a necessity of perceiving this relation, a relation perceived but dimly if at all by the mere animal. This is the power which leads the child to ask *why*, and the man to say *therefore*. It is the power which frames syllogisms, and is compelled to accept the conclusion if the premises be granted. It is the power which leads the farmer when he sees a pile of upheaved earth in his garden to conclude that a mole has burrowed there ; which induces the explorer when he discovers the earth-mounds in Ohio or the cliff-dwellings in Colorado to conclude that man has been there before him ; which compels men every-where, seeing the marvelous mechanism of nature by which he is surrounded and in which he dwells, to be sure that some First Great Cause has called it forth. This is the power which guides man in his search, whether it be the search of the farmer for the hiding mole, the antiquarian for the lost race, or the philosopher for an unknown God. It is this power which enables us to trace sequence in nature, in history, in human experience ; which enables us to see that phenomena are not isolated and accidental, but every fact is a link in an endless chain. The exercise of this faculty upon numbers gives us the higher mathematics, exercised upon visible phenomena it produces science, upon mental experience it creates history, political economy, mental and moral philosophy. Every mechanic relies upon it when he builds his engine or constructs his dam, sure that the same cause

will always produce the same effect; every artist depends
upon it when he mixes his colors, certain that the same mix-
ture will produce the same shade, and the same touch of the
brush on the canvas will cause the same effect; every orator
assumes it, consciously or unconsciously, when by the utter-
ance of his own emotions he strives to awaken the emotions
of his hearers, or by the process of reasoning he expects to
win the assent of their judgment to his own conclusions. It
is, indeed, one of the master powers of man's mind, but it is
not the master power; and when men attempt to make it do
the work of the supersensuous faculty, when they ignore the
power given them to perceive directly and immediately the
invisible world, and attempt in lieu of exercising that faith-
power with which they are endowed to arrive at the truth
respecting the invisible world by employing the logical faculty upon observed phenomena, the result is always a ration-
alism, which, whether its conclusions be orthodox or hetero-
dox, those of a Bishop Butler or those of a John Stuart Mill, is
far removed from that spiritual religion which is founded on
experience, not on deduction, and which says not, I conclude,
I think, or I believe, but I know.

2. But while the relation of cause and effect is the relation
which binds phenomena together, there are other relations
than those of cause and effect which the mind must perceive
in order that it may classify phenomena, and truly apprehend
their meaning and value. The phrenologist calls this faculty
comparison, a name which seems to give to it too limited a
scope; and yet it is not easy to suggest for popular use a better title. By this power, called by whatever name, man perceives both differences and resemblances. He perceives the
vital difference between a whale and other citizens of the
ocean, and discovers that a whale is not a fish; he perceives
the resemblance between the spark of electricity and the
thunder-bolt, and out of this perception grows all electrical
science; he perceives the resemblance between the fall of

an apple and the movement of the earth, and out of this perception grows the whole science of modern astronomy ; he perceives the resemblance between man and the animal, and out of this perception grows comparative physiology. All science is based on this power; all eminent scientists possess it in an eminent degree, and employ it continuously, and often almost unconsciously.

But this is by no means its only, perhaps not even its chief, scope. It traces resemblances between the outer and the inner world, the visible and the invisible. It is the poet's brush and the orator's, whereby they cast upon their canvas a thousand tropes and figures and metaphors. All language of the inward life employs unconsciously this subtle power of perceiving analogies. " He is frozen with horror; " " he is full of wrath; " " he is struck with an idea: " these and kindred phrases in our daily conversation are all based upon the possession of a power in man to see the subtle analogies between spiritual experience and external phenomena. These analogies make all life a parable ; this power makes every man in some measure a poet and a prophet. It enters largely into all imagination, which is sometimes indeed the direct perception of invisible realities, but which is sometimes also the construction of new images by the power which perceives relations before unperceived, and brings together objects familar in forms and combinations before unknown. The ancient centaur may be fairly taken to illustrate both types of imagination. By his supersensuous faculty, his faith-power, the poet saw in the soul of man the strange amalgam between the bestial and the divine ; this was no visible disclosure, no logical deduction ; it was a spiritual perception. To embody it to the senses of others, he combined the head and breast of a man with the body of a horse, an unreal combination of real things, to illustrate a real but invisible combination. The faculty which perceived the invisible amalgam was one, the faith faculty. The faculty which framed the visible combination was another, the faculty of comparison. The product of the two we call the product of the imagination.

This same faculty lies at the basis of all wit and humor. Two of the most difficult problems of mental science are: What is the secret of the beautiful? What is the secret of the humorous? I shall not enter here into these old problems, still less attempt to solve them. It is enough to say that by a general, if not universal, consent the foundation of both wit and humor is a sudden and unexpected discovery of either a disparity or a resemblance. Other elements enter into it. Not every such unexpected and sudden discovery produces a tendency to laugh. But it may safely be said that the faculty of comparison is always called into play in every ebullition of wit and humor, and that those whose faculty of comparison is either feeble or slow to act are never quick to take a joke, and rarely greatly enjoy it.

CHAPTER XI.

ATTENTION, MEMORY, WILL.

THERE are three other mental powers which are sometimes treated as separate faculties, *attention, memory,* and *will.* I do not so treat them here for the same reason that I have not treated hope and firmness as separate motive powers. Attention and memory are rather mental habits than mental faculties, and will is the power which reigns over all the faculties; it is the personality, the individuality, which, so to speak, administers the whole kingdom.

1. Attention is a habit or power of concentration, which may characterize one faculty or another, or all combined. It is, however, usually a concentration for the time of all the soul's powers upon a single faculty, and is generally proportioned to earnestness of desire. The merchant finds no difficulty in concentrating attention upon the business of his counting-room ; he does it without conscious effort ; but when he has returned home in the evening, it is with the greatest difficulty that he concentrates it upon a book, and sometimes reads a page before he discovers that his mind has been upon the business of the day, not upon the Shakespeare in his hand. So a college boy easily puts his whole mind upon a ball game, but requires a vigorous act of the will to fasten it upon his Cicero or his Homer. The secret of attention is interest, and when the faculty is aroused by a strong motive all the power of the soul is concentrated on the problem before it instantly and instinctively and without an effort. The mother who is praying for the recovery of her sick child does not find herself troubled by the wandering thoughts in prayer which have been the bane of her public worship so frequently in church.

2. As attention is the concentration of the faculties upon a subject, induced by strong interest in it, so memory is the retroactive action of each faculty. It is not a separate faculty, as though the powers of the mind gathered truth and the memory received and stored it. It is generally proportioned both to the strength of the particular faculty and the interest in the particular subject. A man who has a well-developed faculty of numbers will remember dates; a man who has a well-developed faculty of color will remember the picture which another has forgotten. One mind will remember facts and principles, another words and localities. Mr. Maurice mentions as extraordinary the memory of his sister for the fragrance of a particular flower inhaled in her childhood; but his memory for principles is evident on every page of his published works. Joseph Cook will repeat with almost verbal accuracy a paragraph read months, perhaps years, before. It is said of Henry Ward Beecher that, during the revival of 1856–7, at a prayer-meeting in Burton's Old Theater, he declined to lead the congregation in the Lord's Prayer, because he dared not trust himself to repeat it without the book. For myself, I am never able to cite an author, or quote a text of Scripture, or a verse of a familiar poem, with any assurance of accuracy; but I can go to my library, take down the book where I have seen or read the sentence I wish to quote, and turn to it, generally at once, though years have elapsed since I saw it. In another edition, differently paged, I might search for it in vain. This simple illustration may suffice to make clear my meaning, if not to demonstrate its accuracy, that memory is simply the power of a faculty to retain what it has once acquired, or repeat what it has once done; a power which depends usually, if not always, upon the degree of interest which attached to the first acquisition, or upon the force which attached to the first action.

3. Into the question, the most hotly debated of all ques-

tions in mental science, of the freedom of the will, neither the limits of this little treatise, nor my object in writing it, allow me to enter. But certainly any study of human nature would be fatally defective which failed to afford any philosophical statement concerning the will; nor can such a statement be made and the question at issue between the two great schools, both in philosophy and theology, be ignored. I must content myself with very briefly stating the difference and my own conviction, without entering into any argument in support of the one position, or into any criticism of the other.

One school of philosophy holds that the motive powers heretofore described are the ultimate facts of human nature; that man is made up of these powers; that he is and necessarily must be governed by the strongest of these motives. The argument is very simple and not easy to answer. Man must be governed by the strongest motive; for if not, then he would be governed by some other motive, which would, therefore, be stronger than the strongest, which is absurd. This is the philosophy, psychologically stated, of what is known as the Calvinistic school of theology, though repudiated now by many, perhaps by a majority of modern Calvinists. It is the theory also of a large school of modern scientists, who hold that mental phenomena are as truly subject to undeviating law as physical phenomena, and that every thought and emotion, no less than every force and phase of nature, is one link in an endless and never-to-be-broken or ended chain. In this view the will is only the balance of the faculties and the preponderance of the strongest. Philosophically it leaves man the creature of an inscrutable fate. Religiously this condition is escaped, because religion points to a God who is able, by a direct and supernatural intervention, to make strong the conscience and the love, which are by nature weak; a God who will always thus interpose to save the lost from his own undoing, whenever, overpowered by a sense of his hopelessness, he appeals to his Creator and Redeemer for that divine help without which he never can choose the good or turn away from the evil.

The other school of philosophy holds that man himself is more than all the motive powers within him; that he possesses what has been called a self-originating power of the will; that his will, that is, he himself, his personality, the *ego* which makes him a free moral agent, has power not only over things external to him, but over his own appetites, desires, inclinations, and is able to curb the one, and directly or indirectly to quicken and strengthen the other; that he dominates himself; that he is not like a chip, the prey of every wind or wave, nor like a steam-ship, controlled by its own sails and its own engines; but like the same steam-ship when sails and engines are controlled by a master, who uses them to accomplish successfully his predetermined voyage.

Which of these views of human nature will be taken by the student of life and character will be determined largely by the question whether he looks upon human nature from the outside or the inside. If he observes what it appears to do when studied by the "scientific method," or determines what it must be presumed to do from considerations derived from theories of man's nature, God's nature, and the divine government, he will tend toward the Necessarian theory of life. If, on the other hand, he looks within, takes the testimony of his own consciousness and that of others, and believes the witness which men bear to their own interior conviction of freedom, he will tend toward the other view. Samuel Johnson expressed this contrast by his saying, "All argument is against the freedom of the will; we know we're free, and that's the end on 't." All mere external observation and all *a priori* reasoning respecting human nature I regard as not worthy to be compared with the testimony of consciousness. It is the universal testimony of consciousness that there is a freedom of will, a power superior to the motive powers, a real self control, an *ego* which is not controlled by, but itself controls, every inward impulse and every intellectual power of the soul and spirit. This *ego*, this master of the moral mechanism, is the will; and in its last analysis all moral action and all moral character depends on the action and on the character of this *ego*, this master of the whole nature, this captain of the ship, his lord of the intellectual and mental domain.

www.ingramcontent.com/pod-product-compliance
Lightning Source LLC
Chambersburg PA
CBHW022145090426
42742CB00010B/1406